This book belongs to

_____

_____

_____

銀色騏驥
小
繪

Color Test Page

Dear Customer

A thousand thanks for purchasing this book, which motivated us to continue providing outstanding products. We really appreciate your choice.

We are a young publisher but we have big hearts and a big vision. We do our best to offer the highest quality books for you to enjoy. If you enjoy this book, please take a few seconds to share your exciting experience on the Amazon product page. This helps us keep improving our products, and help potential buyers to make confident decisions. We wish you a meaningful experience with this book. Thank you, again.

Sincerely

Modern Art Publications

Credits

Made in United States
Troutdale, OR
07/28/2024